STRONG KIDS HEALTHY PLATE

3

Sensible Salts, Sugars, and Fats

Katie Marsico

Cherry Lake Press

Published in the United States of America
by Cherry Lake Publishing Group
Ann Arbor, Michigan
www.cherrylakepublishing.com

Content Adviser: Debbie Fetter, PhD, Assistant Professor of Teaching Nutrition, University of California, Davis
Reading Adviser: Marla Conn, MS, Ed, Literacy specialist, Read-Ability, Inc.

Photo Credits: ©Singkham/Shutterstock.com, front cover; ©Atsushi Hirao/ Shutterstock.com, 1; ©Marian Fil/Shutterstock.com, 4; ©YAKOBCHUK VIACHESLAV/Shutterstock.com, 6; ©Prostock-studio/Shutterstock.com, 8; ©zhukovvvladg/Shutterstock.com, 10; ©fizkesg/Shutterstock.com, 12; ©Evan Lorne/Shutterstock.com, 14; ©Nancy Hixson/Shutterstock.com, 16; ©urbans/ Shutterstock.com, 18; ©Sergey Novikov/Shutterstock.com, 20

Cherry Lake Press is an imprint of Cherry Lake Publishing Group.

Library of Congress Cataloging-in-Publication Data

Names: Marsico, Katie, 1980- author.
Title: Sensible salts, sugars, and fats / written by Katie Marsico.
Description: Ann Arbor, Michigan : Cherry Lake Publishing, 2020. | Series: Strong kids healthy plates | Includes index. | Audience: Grades K-1 | Summary: “Fat, salt, and sugar are a flavor food, not a food group. Learn how much of these flavor foods you should eat to stay healthy and maintain a healthy diet. Content encourages balance and making healthy choices. This level 3 guided reader is based on the U.S. government’s diet recommendations. Readers will develop word recognition and reading skills while learning about food and where it comes from. Includes table of contents, glossary, index, author biographies, and word list for home and school connection”— Provided by publisher.
Identifiers: LCCN 2020002841 (print) | LCCN 2020002842 (ebook) | ISBN 9781534168671 (hardcover) | ISBN 9781534170353 (paperback) | ISBN 9781534172197 (pdf) | ISBN 9781534174030 (ebook)
Subjects: LCSH: Nutrition—Juvenile literature.
Classification: LCC QP141 .M267 2020 (print) | LCC QP141 (ebook) | DDC 612.3—dc23
LC record available at https://lccn.loc.gov/2020002841
LC ebook record available at https://lccn.loc.gov/2020002842

Cherry Lake Publishing Group would like to acknowledge the work of the Partnership for 21st Century Learning, a Network of Battelle for Kids. Please visit http://www.battelleforkids.org/networks/p21 for more information.

Printed in the United States of America
Corporate Graphics

Table of Contents

About the Author

Katie Marsico is an author of nonfiction books for children and young adults. She lives outside of Chicago, Illinois, with her husband and children.

What do you need energy for?

A Limited Part of Your Plate

Want a snack? Try low-fat yogurt or trail mix. Eat a bowl of fresh fruit.

Have a tasty treat. But **limit** the fat, salt, and sugar!

Fat, salt, and sugar **flavor** food. They help it last longer.

Yet fat, salt, and sugar aren't in any **food group**. They aren't meant to be a big part of your **balanced diet**.

The Facts About Fat

Some fats are made from animal **products**. They're often used in desserts and fried foods.

Most fats made from **oil** are better for you. These oils come from fish or plants.

Healthy fat choices give you **vitamin E**. Five teaspoons (25 milliliters) of oil a day, or less, are good for you. You're probably already getting this. A 5-teaspoon (25 ml) serving is about a handful of almonds.

A small amount of fat and oils from plants is healthy for you.

Too Salty?

Salt adds **sodium** to your diet. Your body uses sodium. But you only need a little. Too much sodium is bad for your heart.

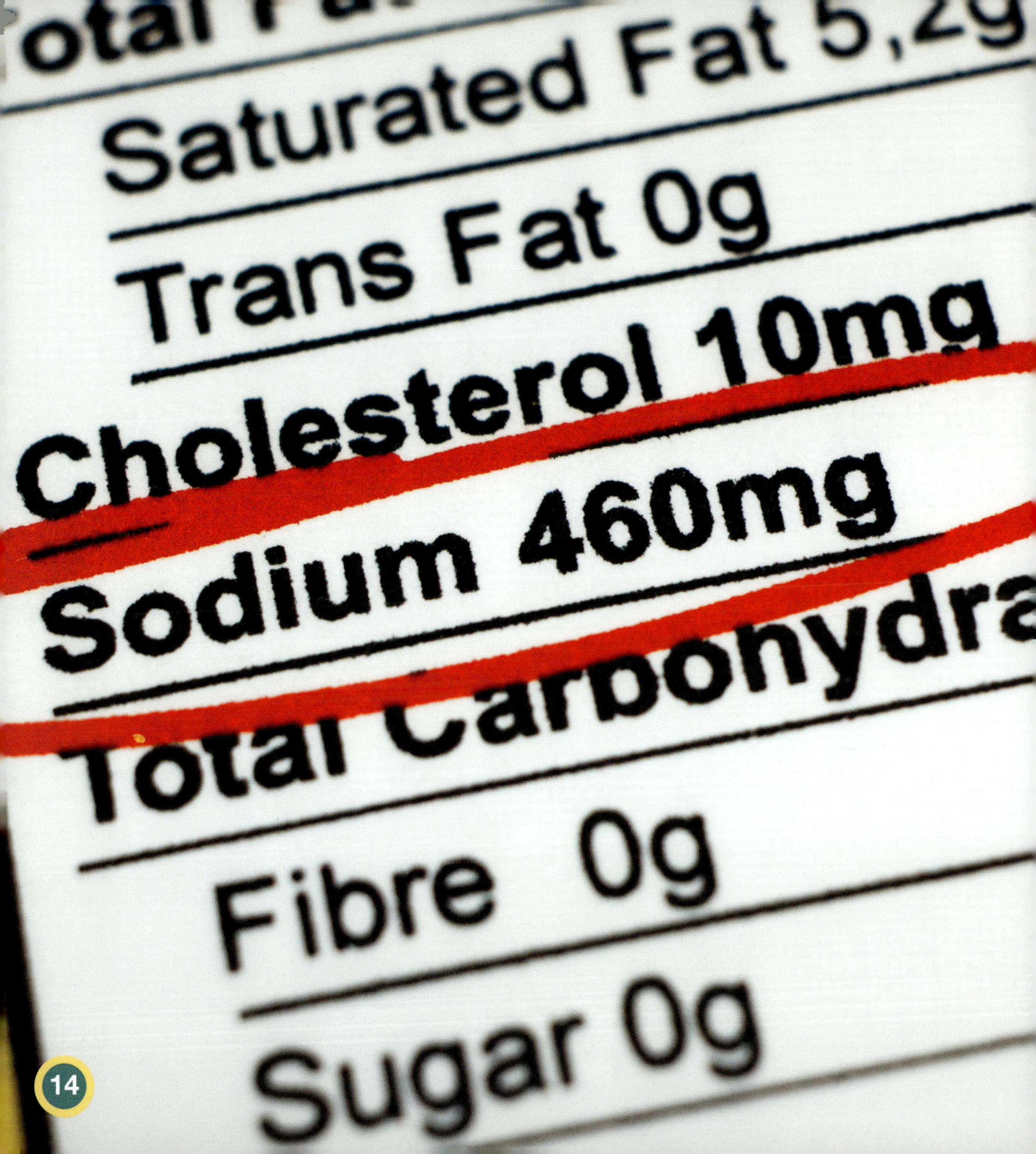
Saturated Fat 5,2g
Trans Fat 0g
Cholesterol 10mg
Sodium 460mg
Total Carbohydra
Fibre 0g
Sugar 0g

Limit yourself to 1 teaspoon (5 ml) of salt a day. It helps to read labels. Watch for added salts. How much sodium is in your favorite foods?

Switch Sweets!

A little sugar gives you energy. **Natural** sugar is best though. It's found in milk and fresh fruit. It's also in 100 percent fruit juice.

Other sugars might be added to foods. These sugars sweeten candy and soft drinks. Limit added sugars. They make it hard to stay at a healthy weight.

Being Balanced

Is it awful to eat cake or fries? No! Treats are okay now and then.

You don't have to give up fat, salt, and sugar. Simply set limits and balance your diet!

Glossary

balanced diet (BAL-uhnsd DYE-it) eating just the right amounts of different foods

energy (EN-ur-jee) the ability to be mentally and physically active

flavor (FLAY-vur) to improve or change how something tastes

food group (FOOD GROOP) one of five groups of different foods that people should have in their diet

limit (LIM-it) to set a certain amount that you should not go beyond

natural (NAH-chuh-ruhl) found in or made by nature instead of by people

oil (OYL) a more liquid form of fat that often comes from plants or fish

products (PRAH-duhkts) things that are created and prepared for sale

sodium (SOH-dee-uhm) a substance that affects how your body controls muscles, nerves, and fluid levels

vitamin E (VYE-tuh-min EE) a substance found in good fats that supports eye, heart, and blood health

Home and School Connection

a
about
added
adds
almonds
already
also
amount
and
animal
any
are
aren't
at
awful
bad
balance
balanced
be
being
best
better
big
body
bowl
but
cake
candy
choices
come
day
desserts
diet
do
don't
drinks
eat
energy
facts
fat
fats
favorite
fish
five
flavor
food
for
found
fresh
fried
fries
from
fruit
getting
give
good
group
handful
hard
have
healthy
heart
help
helps
how
in
is
it
it's
juice
labels
last
less
limit
limited
little
longer
low
made
make
meant
might
milk
mix
most
much
natural
need
no
now
of
often
oil
oils
okay
one
only
or
other
part
percent
plants
plate
probably
products
read
salt
salty
serving
set
simply
small
snack
sodium
soft
some
stay
sugar
sweeten
sweets
switch
tasty
teaspoon
the
then
these
they
they're
this
though
to
too
trail
treat
try
up
used
uses
vitamin
want
watch
weight
what
yet
yogurt
you
your
yourself
you're

Find Out More

Book

Reinke, Beth Bence. *Healthy Eating Habits*. Minneapolis, MN: Lerner Publications, 2018.

Website

USDA—MyPlate Kids' Place

www.choosemyplate.gov/browse-by-audience/view-all-audiences/children/kids

Use online learning tools to be healthy and build a more balanced diet.

Index